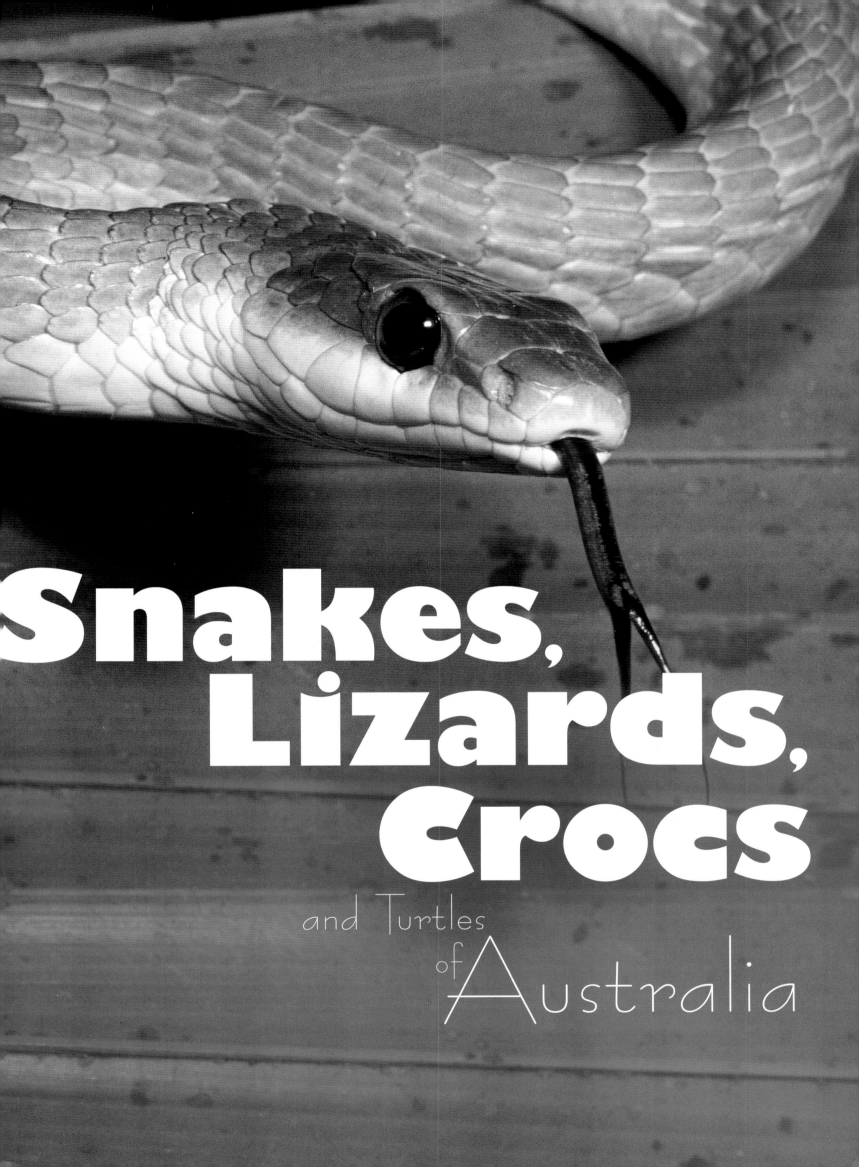

Snakes, Lizards, Crocs

and Turtles of Australia

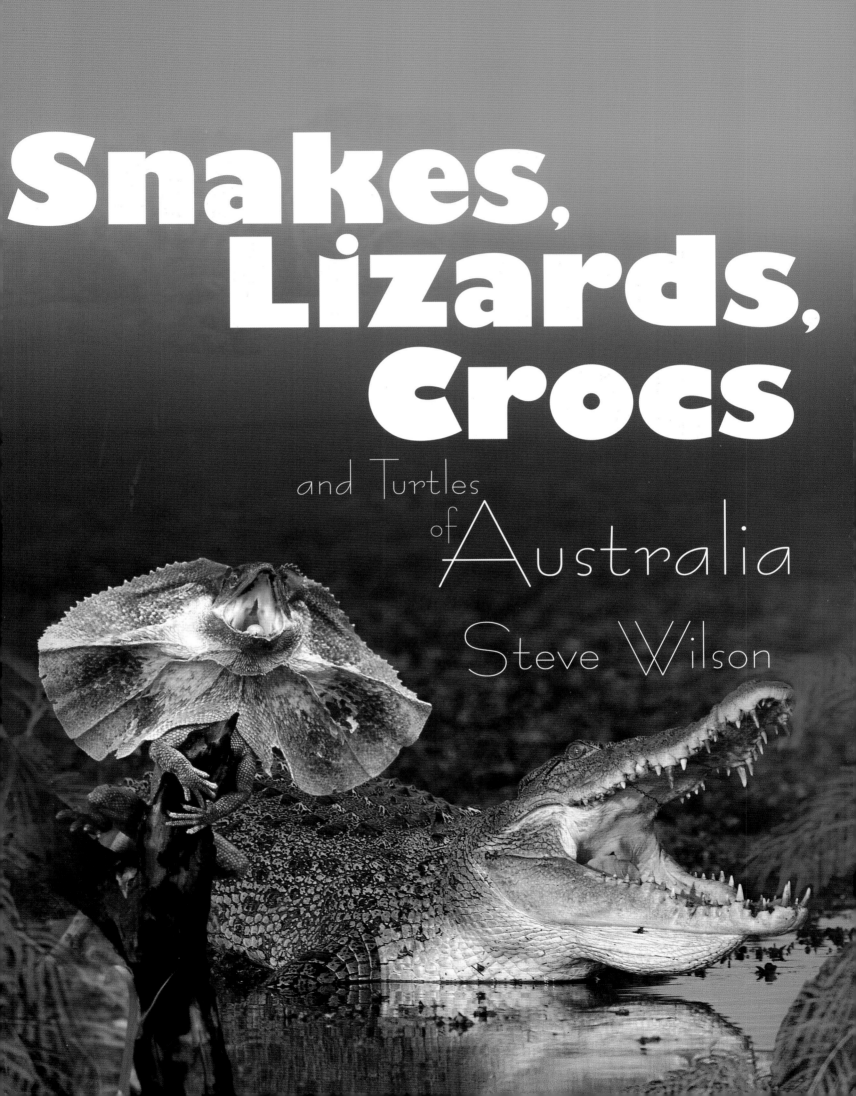

Snakes, Lizards, Crocs

and Turtles of Australia

Steve Wilson

How to Use this Book

This book is full of information about Australian reptiles. The first section of the book, 'Meet the Reptiles', introduces you to all the different kinds of reptiles. Each double-page spread that follows shows a different type of habitat and the animals that live there.

The map on the next page shows most of the habitats covered in the book. Each coloured box matches the habitat listed beside it in the contents list. You can then find the habitat on the map.

Note that habitats usually merge gradually into each other, but they have been drawn on the map with sharp edges to make them clearer. Habitats are also usually quite mixed, with rocky hills rising from open plains, gum trees lining the banks of desert waterways, and rainforest pockets growing in gullies across northern Australia. This map does not cover such details, but it will give you a general idea of the main Australian reptile habitats.

The 'Going...Going...' section tells you about some of the reptiles that are endangered or may already be extinct.

The 'Activities' section includes Pick 'n' Match, which tests your skills to match reptiles with their scales and colours. Then see how carefully you've been reading with The Reptile Quiz. The answers are all in the book.

The names of the animals are printed in bold so they're easy to find. Other bold words are explained in the Glossary on pages 42–44. There's also an Index and a list of other books and websites you can go to if you want more information.

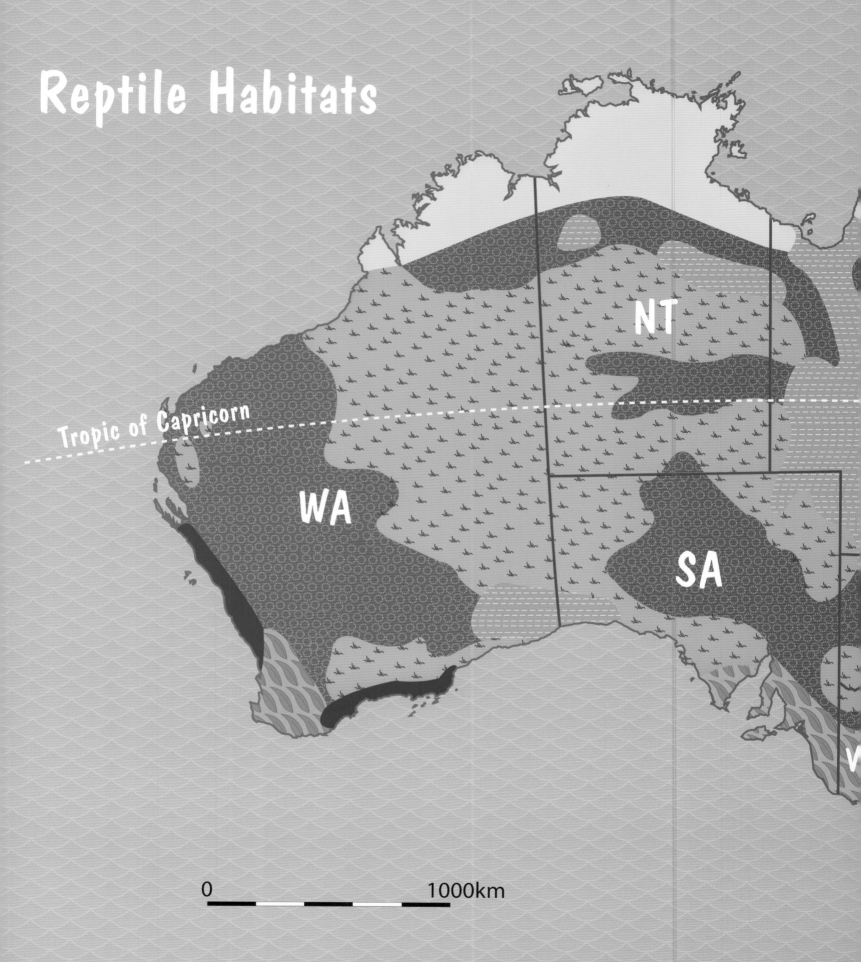

Reptile Habitats

Tropic of Capricorn

NT

WA

SA

0   1000km

Contents

QLD

NSW

TAS

Meet the Reptiles

Australia has a greater variety of reptiles than most other places in the world—more than 800 different **species**. From the burning hot desert sands to the windswept, snowy Tasmanian mountains, and even in your garden and beside the local creek, reptiles thrive in our most extreme environments. Reptiles have backbones and scale-covered skin. They lay eggs on land, or give birth to live young that look like small versions of their parents. And they are 'cold-blooded': they can't make heat like us, and they need to **bask** in the sun or lie on warm surfaces like rocks to get warm. Australian reptiles include crocodiles, turtles, lizards and snakes.

Crocodiles

These ancient armour-plated reptiles have been lazing in the waters of our hot, **tropical** north and snapping at victims with their toothy jaws for millions of years. Some would have dined on dinosaurs when they had the chance!

*Freshwater Crocodiles are generally harmless, but this **Saltwater** or **Estuarine Crocodile** eats people. Watch where you swim!*

Turtles

Turtles are all protected by a tough, bony shell and their limbs are either flipper-shaped or webbed for swimming. Different kinds of turtles **inhabit** salt and fresh water. Sea turtles have flippers and spend their whole lives at sea— except for the females, which must haul themselves onto beaches to lay eggs. Most freshwater turtles have clawed, webbed feet and live in creeks, rivers, swamps and ponds. Some are **vegetarians** while others eat shellfish, frogs and insects. Some are long-necked and some are short-necked but all fold their necks sideways into their shells.

The freshwater **Saw-shelled Turtle** eats both animals and plants. It can even swallow young Cane Toads, which are poisonous to most predators.

Tortoise or Turtle?

Although freshwater turtles are sometimes called 'tortoises', we do not have any tortoises in Australia. True tortoises live only on land. All our turtles are **aquatic.**

This scientist has caught a **Loggerhead Turtle**, which is a sea turtle. Before releasing it he will weigh, measure and tag it. If it is captured again, we can learn about its growth, age and migration.

Pig-nosed Turtle

This unusual turtle lives in clear freshwater rivers in the north of the Northern Territory. It is unrelated to any other Australian turtles. It has both claws and flippers and a soft covering over its shell. It also withdraws its head directly backwards rather than folding it sideways like other freshwater turtles.

Pig-nosed Turtles have a short, fleshy trunk for a nose. But why? No-one knows.

Flap-footed Lizards

Flap-footed lizards appear very snake-like. They have no front limbs and only small scaly flaps for hind limbs. Like the geckos, flap-footed lizards can squeak if harassed and can easily break and regrow their long tails. They also clean their lidless eyes with their broad tongues.

*It is easy to mistake this **Common Scaly-foot** for a snake, but it has ear-openings and a very long tail.*

Dragons

Dragons don't breathe fire, but they do have rough skin, long tails and alert, upright postures. Some have impressive rows of spines down their backs. Others can erect frills or beards when alarmed. Most dragons are very swift, and some can even run upright on their back legs.

Tail Tales

Lots of lizards can break off their tail if someone grabs them. The broken tail wriggles around creating a distraction while the lizard escapes. And the lizard grows a new tail.

Geckos

Geckos are **nocturnal** lizards with soft skin and large, lidless eyes. Some **species** have padded fingers and toes with special microscopic bristles that grip smooth surfaces such as walls, windows and even ceilings.

*Geckos cannot blink. This **Smooth Knob-tailed Gecko** uses its flat, wet tongue like a windscreen-wiper to clean the clear covers over its eyes.*

*The bright colours of this **Tawny Dragon** from South Australia show that it is a male.*

*This **Eastern Striped Skink** has lost its tail and the damaged stump has sprouted two new tails!*

Snake-skinks *have tiny eyes, no visible legs and no ear-openings. They burrow into soil and compost.*

__Gidgee Skinks__ arch their backs and use their spines to wedge themselves into cracks so predators cannot pull them out.

Skinks

Typical skinks are the small, quick lizards in your garden. They usually have shiny, overlapping scales on their bodies, and neatly arranged plate-like scales on their heads. Most have four legs, but some burrowing varieties have partly or completely lost their legs and look worm-like. Others, such as blue-tongues, are slow moving and grow to nearly half a metre long.

Lizards

Lizards are the most common reptiles in our bushland and gardens. Most have four obvious legs but others have tiny legs or have completely lost their legs and look more like snakes. These burrow in soil or wriggle through thick low vegetation and sometimes it takes a keen eye to spot the difference! All Australian lizards (except goannas) have flat, fleshy tongues, quite unlike the thin, deeply forked tongues of snakes. Depending on their size, lizards eat a variety of large and small animals. Goannas can rip flesh from dead wallabies, Thorny Devils catch tiny ants, and skinks and geckos hunt small insects and spiders. Many lizards also eat fruit, nectar, flowers and even tree sap.

Spotted Tree Monitors *are small tree-climbing goannas that live in hollow limbs and beneath loose bark across northern Australia.*

Goannas

Goannas, also called monitors, have tough loose skin, strong claws, long tails and long, deeply forked tongues like those of snakes. They include the world's largest lizards, with some growing to several metres long. Goannas are **predators** and **scavengers**. They have long, sharp backward curved teeth.

*It would be easy to mistake this harmless **Blackish Blind Snake** for a worm. Its head is clearly visible but hard to recognise because scales cover the eyes.*

Blind Snakes

These burrowing snakes have shiny scales and blunt tails tipped with a tiny spur. Their small eyes are covered by scales. They cannot see shapes but are sensitive to light. They feed on termites and the eggs and **larvae** of ants.

*It will take weeks for this **Carpet Python** to digest a small wallaby. The snake will not eat again for several months.*

Pythons

Pythons include the world's largest snakes. They have no **venom**, but are well armed with long, curved teeth and powerful bodies. Pythons kill their prey by squeezing it in their coils. Thanks to their flexible skulls and elastic skin they can swallow large prey.

*By flickering its forked tongue, this **Golden Tree Snake** is investigating potential danger— a man with a camera!*

*Underwater this **Arafura File Snake** is a graceful swimmer, but on land it is slow, clumsy and looks like a wet sock.*

File Snakes

There are only two **species** of file snake in Australia. Both live in tropical rivers, swamps and **estuaries**. They are sluggish, **non-venomous** and feed on fish. File snakes have loose, baggy skin that is rough like a file.

Tongue Talk

Snakes and goannas use their long, deeply forked tongues to find food and mates, and to sense danger. Their tongues 'smell' by collecting tiny particles (the same ones dogs can smell with their noses). And because the tongues are forked, they can pinpoint direction.

Elapid Snakes

Australian **venomous** snakes belong to a group called elapids. They have fangs at the front of the mouth connected to a **venom gland**. Most are not dangerous, but some, such as taipans, brown snakes, tiger snakes and death adders, are among the world's most deadly snakes. Most live on land but some, with flat paddle-shaped tails, live in the sea.

Northern Death Adders *hide under leaves, hoping to* **ambush** *birds and lizards. Their* **venom** *is deadly so watch your step!*

The **Black-striped Snake** *is less than 30 centimetres long. It is harmless and lives only in sandy areas near Perth.*

Snakes

Snakes live everywhere in Australia, even in the sea. They are closely related to lizards but all have distinctive forked tongues and lack the ear-openings that most lizards have. Snakes eat only animals—other reptiles, frogs, mammals and birds—and they normally swallow them whole. Ancestors of snakes had limbs. Pythons still retain their back legs in the form of tiny spurs. This means that some snakes have more legs than some lizards!

Colubrid Snakes

These are the tree snakes, freshwater snakes and **mangrove** snakes. There are only 10 **species** in Australia. Some are **non-venomous**, others are weakly **venomous** with fangs at the backs of their mouths. Most live in **tropical** areas, though tree snakes are found south of Sydney.

The poison oozing from this **Cane Toad** *will not save it.* **Keelback Snakes** *can often swallow toads and survive. Many of their relatives in other countries thrive on toads!*

Face Full of Food

Snakes can swallow large prey because they have very flexible skulls and their skin can stretch. You can only move your lower jaw, but snakes can move both sides of the upper and lower jaws separately. By shifting the upper jaw forward on one side, then the lower, and repeating the process on the other side, they can basically 'walk their faces' over their dinner.

*This **Curl Snake** has caught a mouse and is eating it whole. Curl Snakes are **venomous**. They live in dry areas over much of Australia and hide in deep soil cracks.*

Fussy Eater

Like all geckos, the **Pretty Gecko** eats insects, but its favourite food is termites. During the heat of the day it hides down a spider's burrow but now this Pretty Gecko is out hunting.

Termite Treats

Some scientists believe that Australia has such a huge number of desert lizard **species** because there are so many termites for them to eat.

Odd Python

Most pythons eat mammals and birds, but **Womas** eat mainly reptiles. They are not good climbers either. While other pythons are equally happy on trees, rock faces or the ground, Womas live in burrows. Because so much of their **habitat** has been cleared, there are fewer Womas around. They have virtually disappeared from parts of the south-west.

Giant Lizard

At nearly 2.5 metres long, **Perenties** are our largest lizards and the biggest native predators on land in Australia. They can swallow large **venomous** snakes and rip the guts out of dead wallabies! If attacked, Perenties can also use their muscular tails as a lash.

*With its long limbs and toes, the **Seven-lined Striped Skink** is built for speed.*

Desert Speedster

Lightning-fast striped skinks thrive in dry areas across Australia. There are more than 90 different **species** and many live only in deserts. Some are secretive, never moving far from low bushes and grasses, while others often patrol the open spaces.

Living Pebble

It looks like two stones and a twig, but there's actually a **Pebble Dragon** hiding among the rocks. As long as it stays still, few predators will spot this clever pebble mimic.

Scorched Earth

Australia is one of the world's driest continents. Ours is a land of baked red earth, vast plains of wind-polished stones, weathered ancient mountains and rivers that lie **parched** and empty for years on end. There are few people over huge areas of Australia. But there are plenty of reptiles! In fact, most kinds of Australian reptiles thrive in **arid** conditions.

In the Hot Seat

You could just about fry an egg on the rock, yet a **Yinnietharra Rock Dragon** can survive the midday desert heat. While other reptiles have headed for shelter, it avoids being cooked by perching on its heels high off the hot rock. But occasionally it dashes into the shade to cool off. This rock dragon lives only on one remote cattle station in the middle of Western Australia.

Sun, Sand and Spinifex

At midday in the desert it's hot enough to grill a gecko, so most lizards and snakes must hide away in cool, humid burrows or in dense, spiny **spinifex** clumps. In the morning, late afternoon and at night, they come out to look for food. At dawn the sun's sharply angled rays highlight the tracks they leave behind in the soft red sand—tracks that tell tales to anyone who can read them.

Spines and Spikes

With short jerky steps like a clockwork toy, **Thorny Devils** win no races. But few predators could swallow this prickly mouthful. Thorny Devils eat only ants. Lots of them. With dabs of their short, sticky tongues they can gobble up to 5000 or more in a meal.

Sneaky Snake

Desert Death Adders like to hide beneath **spinifex** or under dry leaves at the base of a shrub. They lure unwary prey by wriggling their thin tails to copy caterpillars. Death adders may not move for days at a time, but can strike with lightning speed to catch a passing lizard or mouse.

*The **Jewelled Gecko** rarely leaves its home in dense, spiny spinifex tussocks.*

Goo-squirting Gecko

When threatened, the **Jewelled Gecko** oozes gross sticky goo from glands in its back and tail! With its grasping, padded **digits**, it climbs easily among the tough, slender **spinifex** needles where it lives.

Fierce Predator

Huge **Yellow-spotted Goannas** roam the sand dunes, seizing prey with their long, sharp backwards curved teeth. They use their long, deeply forked tongues to follow scents and their powerful claws to excavate burrows.

Yellow-spotted Goannas grow to more than a metre long and they don't seem to care how hot it gets in the desert.

Greased Lightning

Quick as a flash, **Military Dragons** sprint across open sandy patches on their long legs, pausing only to snap up an ant or keep an eye out for danger. A host of predators would love a tasty feed of lizard. If only they could catch one!

Spiders for Dinner?

The **Knob-tailed gecko** thrives in the harsh desert environment, eating scorpions, spiders and even other geckos. It is **nocturnal**, patrolling cool sand that, only a few hours before, would have roasted it within seconds.

Underground Dramas

Slider skinks have wedge-shaped noses and streamlined bodies. They have even lost some limbs and digits and so they can swim in loose sand like fish in water. **Keeled Sliders** (left) leave tracks like wandering squiggles across the desert sand.

*The venomous **Black-naped Snake** wriggles just beneath the soft sand, searching for its favourite food—slider skinks!*

*The Smooth **Knob-tailed Gecko** has soft skin, big lidless eyes and a frail-looking body.*

The broad flat tail, prickly skin and dappled pattern provide perfect camouflage for this gecko.

This snake is an adult. Babies are bright yellow.

One Species in Two Places!

Green Tree Pythons only live in small areas of rainforest on nor[...]
Cape York Peninsula, but are widespread over most of **New Gu[...]**
They live in both places because New Guinea was joined to Aus[...]
thousands of years ago, when sea levels were lower. Now, with h[...]
sea levels, New Guinea is an island and the **populations** are separ[...]

Master of Disguise

Nearly invisible against the mottled bark is a **Leaf-tailed Gecko.** By day it rests in hollow trees but at night it lurks upside down on the tree trunk, waiting to catch insects and spiders.

Hide-and-seek Dragon

Against the dappled light and shifting shapes in north Queensland's **tropical** rainforest, the **Boyd's Forest Dragon** is easy to miss. When approached, it simply slips silently around to the other side of the tree. This one is **basking** in a shaft of sunlight on a tree trunk, scanning the ground for insects and spiders, and watching for rivals or mates.

Nocturnal Hunter

By day **Golden-crowned Snakes** sleep under rocks, logs and thick leaf litter. At night, they come out to hunt, using their forked tongues to pick up the scent of their prey—**diurnal** skinks. They hope to catch the lizards napping after their busy day chasing insects and sunning themselves in puddles of light.

Out of the Egg and Ready for Action!

These eggs were buried in a clearing, where sunshine could reach the ground to warm them. To hatch, the baby **Southern Rainforest Dragons** have cut the soft shells using an egg tooth—a tiny spine on the tip of the snout. Southern Rainforest Dragons live in the **subtropical** rainforests of northern New South Wales and southern Queensland.

Their mother will never return so these young dragons must fend for themselves. It looks like they might do a good job!

Rainforest
Land of Shadows

There are many reptiles in the rainforest, but don't get your hopes up—they can be hard to find! Skinks scuttle and rustle in puddles of sunshine, but all else is quiet. On a thick carpet of soggy brown leaves, logs rot down to a pulp. Giant trees supported by **buttresses** reach skywards for vital sunlight, blocking out most of the light, while vines thicker than your legs twine around their trunks. The branches are **festooned** with ferns and other plants whose roots never touch the ground.

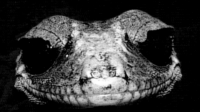

Extraterrestrial

Looking more alien than earthling, a **Chameleon Gecko** looms over a mossy log in north Queensland's Wet Tropics. At night it prowls the forest floor or clings to the stems of thin **saplings**, where it **ambushes** insects. When this gecko breaks its tail, the broken tail not only wriggles but it also squeaks loudly, drawing attention away from the escaping gecko!

Colourful Show-off

In the rainforests of mid-eastern Queensland, **Red and Blue-throated Rainbow Skinks** bask in sunny patches and explore under leaves and in cracks for small insects. They also wave their tails, bob their heads and show off their striking colours. The males are the brightest, but females and the young can also be boldly marked. The colours probably tell who gets to be boss and have more mates.

Cold Comfort

In the **cool-temperate** zone—the highlands of the south-eastern mainland, and the mountains and south-western parts of Tasmania—winters are chilly, and sudden blizzards can even freeze your pants off in summer. There are plenty of reptiles, but not many different **species**. Keeping warm is a big problem for them. Most reptiles in cool climates give birth to live babies rather than laying eggs. There may not be enough warmth in the soil to **incubate** eggs, but if they carry the young inside them until they're fully developed, they can **bask** to provide heat. In sunny patches, sheltered from cold wind, they soak up the sun.

Don't Annoy the Copperhead!

Copperheads have stocky bodies and the scales on their upper lips have pale edges. This is the **Highlands Copperhead** from the Australian Alps. Though dangerously **venomous**, copperheads rarely bite except when severely provoked. Of course, it's never a good idea to annoy any snake.

Copy Cat

Tasmanian She-oak Skinks have long bodies and tails but their legs are very short. They can slither easily through dense low shrubs and grasses. When threatened, they raise their bodies and flicker their tongues. It is a clever trick to make these harmless lizards look like alert snakes, ready to defend themselves.

Small Snake

White-lipped Snakes are the smallest **cool-temperate** snakes, growing to only about 50 centimetres. You'll see them zipping off walking trails into thick tussock grasses.

*Most mainland **Tiger Snakes** have a banded pattern (right), but many Tasmanian ones are plain black (below). This helps them absorb heat more easily.*

Tiger Snakes

Tiger Snakes often hang out near water, where they can catch their favourite food—frogs. Some Bass Strait islands are home to huge Tiger Snakes, up to 2 metres long. They eat plump young muttonbird chicks. Even though these are only around for a short time because they grow up and migrate, there is enough fat and nourishment in the birds to sustain the snakes for the rest of the year.

In the Fridge

Some mountain tops are so cold that there are no trees. **Northern Snow Skinks** (below) live above the **tree line** in Tasmania. For a short time each year they **bask** among rocks, often beside chilly pools and streams. Sometimes they plunge into the icy water and hide under stones to escape danger. In winter, the lizards' body temperatures drop to just above freezing as they hibernate in burrows under rocks beneath a thick blanket of snow.

Cool but Not Cute

All **cool-temperate** snakes are **venomous**. Not all are dangerous, but a bite from even a mildly venomous snake could make you pretty sick, so treat them all with caution.

Is it a snake? No, it's just a harmless flap-footed lizard. But like most small Australian snakes, **Burton's Snake Lizard** eats only lizards. It ambushes them from its hiding place in the grasses, grasping prey in its long, slender jaws. This unusual reptile has a hinged head. This allows the tips of its jaws to meet, enclosing its prey.

Hinged Head

This **Burton's Snake Lizard** *has killed a gecko and will soon swallow it headfirst.*

*This **Frilled Lizard** has its frill folded.*

Famous Frilly

Frilled Lizards are our most famous lizards, thanks to their spectacular 'frills'. They have appeared on coins, postage stamps and book covers. Though common across the Top End, during the Dry Season Frilled Lizards remain in the treetops, rarely showing themselves. When the Wet Season arrives they perch along roadsides and cling to rough-barked tree trunks. Frilled Lizards feed on grasshoppers and other insects, and especially like to eat the millions of swarming winged termites that fly during humid weather.

The Web of Life

Reptiles are much more than just food in Aboriginal society. They are woven into tales of creation and adventure and feature in Aboriginal artwork.

By gaping its mouth and erecting its scaly frill, this **Frilled Lizard** *can make itself look bigger to frighten off predators.*

A Snake With Attitude!

The arrogant-looking and muscular **Coastal Taipan** is one of the world's most lethal snakes. It eats only mammals and has very toxic venom to subdue prey that might bite back. Taipans live in **tropical** dry forests and woodlands but are also at home in cane fields, where there are lots of juicy rats.

Wild Eyes

Sometimes, when animals try to hide, their eyes give them away. This bold pattern probably breaks up the shape of the eyes of this **Northern Spiny-tailed Gecko**, helping it to remain camouflaged as it clings to twigs and foliage. The vertical slit is the pupil, the part it sees through.

Top End
Land of Extremes

During the long, hot Dry Season, the ground cracks in the baking sun, and tall grasses turn brown and shrivel. Fires sweep over the land. Around November, black clouds build up and lightning arcs across the sky. The humidity becomes unbearable. Then the Wet Season starts, and for a few months heavy rain fills the rivers and spreads water across the plains to generate lush green growth.

Look Out!

With mouth agape, a **Night Tiger** strikes. Fortunately, its venom is not strong and its small fangs are at the rear of its mouth so it is not harmful to humans. A skilled climber, it often hunts birds and small mammals along the rocky faces of gorges.

This red-and-white **Night Tiger** *lives in the Top End. In eastern Australia the same* **species** *is patterned more drably, and called a* **Brown Tree Snake**.

Sealed in the Nest

After these brightly coloured youngsters disperse, they will spend their early life high in the trees.

Baby **Lace Monitors** begin life in a termite nest. Their mother ripped into the tough crust, laid her eggs and partly covered them. The soft, near-blind little termites quickly repaired the nest, sealing the eggs in galleries and tunnels where the stable temperature and humidity are perfect for them to **incubate**. Caring for young is rare in lizards, but scientists believe that the mother returns to dig out the babies when they hatch.

Out on a Limb

This striking **Golden-tailed Gecko** lives in dry forests of inland Queensland, where **eucalypts** mix with native pines. Unlike most geckos, which shun the sunlight, this gecko often **basks** in the filtered light and may even cling to slender branches, exposing itself to the hot sun.

Clever Climber

Unlike most **venomous** snakes, the **Pale-headed Snake** is a skilled climber. It spends most of its life in trees, sheltering in the hollows of river gums and behind the loose bark of ironbarks. But it does not climb well on slender branches or foliage, like true tree snakes, preferring to remain on trunks and broader limbs. At night it emerges to hunt frogs.

This snake is cornered and ready to strike. It has powerful venom but no human deaths have been recorded.

Black and White Loops

Sensing danger, a **Bandy Bandy** thrashes, arches and contorts its body into loops, trying to alarm potential predators. Bandy Bandys feed only on worm-like burrowing blind snakes. They may be fussy eaters but they can live in a wide variety of **habitats**, from dry forests to deserts and rainforests.

*In this photo you can clearly see the **Bandy Bandy,** but when it's wriggling through leaf litter, its bold black and white rings merge into grey, helping the little snake vanish.*

Big, Bad Dragon

The rough grey skin of a **Bearded Dragon** provides excellent camouflage, concealing it on logs and stumps, where it often perches motionless for hours. But this one has been discovered, so with prickly beard extended and yellow mouth wide open, it tries to look big and bad.

Gum Trees
Tough Leaves

Every drop of water matters in a dry land. Australia's most familiar trees are the drought-resistant **eucalypts.** They make up the 'sclerophyll' forests, meaning the trees have tough, hard leaves that do not lose too much moisture. Fires rage through these forests regularly, leaving the tree trunks blackened, and triggering new growth. Termites are also at work, hollowing limbs and building nests from the pulped wood. Trees die and weakened trunks and limbs snap and crash to the ground. But the hollows and peeling bark provide cosy **habitats** for the local animals.

Combat

Pitted against each other in a test of strength, two bundles of muscle, claws and jaws fight to be top male. They may look fierce, but serious injuries are rare and the loser will live to fight another day. **Lace Monitors** are common in forests between Victoria and north Queensland. They often hang around picnic and camping grounds, where scraps and the odd sausage make easy pickings.

Sharp Eyes?

Some reptiles have poor vision, but others can see extremely well. Snakes that are active by day often have large eyes. They rely more on vision than most **nocturnal** snakes, which often have smaller eyes and use other senses. Look at this **Yellow-faced Whip Snake**. Do you think it has good eyesight?

Life in the Cracks

Collett's Snakes live only in central Queensland on grassy plains called Mitchell Grass Downs—named after the most common plant. They shelter from heat and hide from predators in deep cracks in the clay soil. The **species** is considered dangerous but no-one is known to have died from its bite. This is because they are secretive and seldom seen.

New Birth

More than two months ago, a female **Collett's Snake** laid up to 14 soft-shelled eggs deep within a soil crack. Now they have begun to hatch. This baby may take more than a day to emerge. It will then shed its skin and begin hunting lizards and small mammals.

Bones for Hearing

Most reptiles seem to have excellent hearing. Because snakes have no visible ears, they were considered to be deaf. Scientists now believe snakes use bones in their skulls and jaws, and even use their body surfaces, to detect ground vibrations and airborne sounds.

Hot Head

The **Black-headed Python** lives in burrows and soil cracks. Its distinctive black head may have something to do with heating. By first revealing only its head, the snake can probably absorb warmth without exposing its body to danger.

Rats Beware!

One of the world's most lethal snakes lives only on remote plains bordering Queensland, South Australia and New South Wales. The fortunes of the **Inland Taipan** are closely linked with Long-haired Rats. During long droughts, rat numbers crash and the snakes go hungry. But following rare floods, **rodents** increase to **plague** proportions. Then the snakes are fat, sleek and full of rats! The snake avoids the rats' sharp teeth by injecting a super deadly dose of venom into one animal. It lets go and follows the rat until it dies. The snake can then safely feed at leisure.

*One bite from an **Inland Taipan** contains enough powerful venom to kill hundreds of thousands of rats.*

Stocky Goanna

The short, stocky **Spencer's Monitor** lives only on treeless Mitchell Grass Downs of Queensland and the Northern Territory. In the shimmering heat-haze it forages widely across the open ground for whatever it can catch or **scavenge,** from **venomous** snakes to grasshoppers. When it gets too hot, the goanna shelters in the deep soil cracks.

Sweeping Plains

No trees break the flat horizons, and the low shrubs or grass tussocks offer little protection during the long, hot summers and sometimes chilly winters. Most plains reptiles shelter in burrows or soil cracks. Exposed grasslands and shrublands cover huge areas of inland Queensland, while the famous Nullarbor Plain (Nullarbor means 'treeless' in Latin) spans the Great Australian Bight between South Australia and Western Australia. There are also smaller, but important, grasslands in New South Wales and Victoria.

Secretive Dragon

Grasslands on the western edge of Melbourne were once home to **Grassland Earless Dragons,** but sadly they have not been seen in Victoria for 40 years. These lizards, named for the scaly covering over their ears, were feared extinct until colonies were found in Canberra and southern New South Wales. The secretive little lizards were overlooked because they hide in spiders' burrows. They have also been discovered on the Darling Downs in Queensland, where they thrive on narrow grassy roadsides and even in nearby cotton and sorghum crops. Because all **populations** are isolated by cleared land, they need to be carefully watched to ensure their numbers do not decline.

Banded Beauty

Jan's Banded Snake burrows under leaves and soft sand. In summer it is well protected from the heat, while in winter it sometimes **basks** about a centimetre below the surface of sun-warmed sand. Like many of our small **venomous** snakes, Jan's Banded Snake eats only small skinks. This tiny snake also uses **constriction**, wrapping its victim in tight coils.

Golden Eyes

By day this **Western Spiny-tailed Gecko** has been clinging to the slender stems of a shrub, its grey body closely hugging the bark. Now it has emerged to catch insects and spiders. It sometimes laps up nectar and sap too.

Thin, Thinner, Gone

If you explore the dense, low heath plants along the lower west coast you might be lucky enough to catch a glimpse of a quick grey blur. The slender, flap-footed **Javelin Legless Lizard** is more like a living, high-speed shoelace than a reptile. It **basks** among the waist-high shrubs but is nearly impossible to approach—it vanishes into the vegetation before you even realise you've seen it.

Only lizards from the south-west have orange flushes on their faces.

The sudden change in pattern on this lizard shows where it has broken its tail and grown a new one.

Spots and Stripes

Common Scaly-foots live in southern and eastern Australia, but only those in the southern **heaths** are brightly painted with lines of bold spots. The markings may help camouflage the lizards among the leaf litter, twigs and dappled shade.

One Skink—Many Names

In South Australia this is called a **Sleepy Lizard,** but Victorians call it a **Stumpy-tail.** In western New South Wales it is a **Boggi,** and throughout much of Australia it's called a **Shingleback.** In the west it's a **Bobtail.** But this is no ordinary skink. Ponderously slow, it would rather display its pink mouth and deep blue tongue than flee. Males and females often mate for life, pairing each spring but spending the rest of the year apart. Females give birth to only one or two very large young.

A Blaze of Colour

Heaths

Spring splashes the coastal south-western **heaths** in red, blue, orange and yellow flowers. The showy colours attract insects or birds to pollinate the flowers of plants such as banksias, unusual kangaroo paws and even shy ground orchids. These diverse, colourful plants thrive in the poor, sandy soil, and among them live many reptiles that are found nowhere else. In particular, many kinds of the snake-like flap-footed lizards live side by side here.

Heath Monitor

Heath Monitors, recognisable by their dark colours and narrow bands, patrol the southern heaths and woodlands. When threatened they sometimes climb trees, but more often they retreat to burrows. Those that have been studied on South Australia's Kangaroo Island are known to lay their eggs in termite nests.

A Fishy Crocodile

With a rapid sideways swipe of its long, slender snout this **Freshwater Crocodile** can catch swift fish. It firmly grips the slippery prey with its fine sharp teeth. Crocodiles have a flap of skin at the back of their throat so they do not swallow water when they snap.

A Snake for All Seasons

Water Pythons often live in **tropical** areas that flood each year. During dry times they hunt rats in soil cracks, but when the rains come they live in shallow water, hiding among the reeds and catching the water birds that nest in vast numbers. At Fogg Dam, near Humpty Doo in the Northern Territory, there are hundreds of Water Pythons per hectare.

Backside Breathing

Some freshwater turtles can remain submerged for hours by sucking water into their **cloaca**. They can breathe through their bottoms!

Turtle with Barbels

The fleshy growths on the chin of this **Mary River Turtle** are called barbels. They are believed to help the turtles explore their watery environment. Perhaps the barbels can detect small shifts in the current, or are sensitive to touch in silty water. This short-necked **species** lives in only one south-eastern Queensland river system. It eats shellfish and water plants.

barbels

Stealthy Predator

This **Broad-shelled Long-necked Turtle** is lurking among the soft mud and weeds under a fallen trunk. When a fish or shrimp moves within range, the turtle strikes out with its long, thick neck, opening its mouth to suck in the victim. These turtles are common in south-eastern Australia, but they rarely leave the water and are not often seen.

Painted Turtle

Painted Turtles are named for the colourful splashes of orange on their faces. They share the tropical rivers with many predators, so they need to look out for danger. Birds of prey may swoop down from the air, or crocodiles might snap from the water. At the slightest sign of disturbance this wary turtle will vanish beneath the lily pads.

Water World

A silent, still world lies beneath the green carpet of lily pads. Swift dragonflies skim and turn a few centimetres above the water. Half-submerged branches and trunks break the surface here and there, some ripped from the banks during past floods. But what's that shiny dark spot? Was it there before? Now it's gone! Quietly and without a ripple, a turtle has surfaced for air. Australia's wetlands—our creeks, billabongs, lakes and swamps—provide food, shelter and safety from predators to an assortment of **aquatic** reptiles. Many others live among the river gums, paperbarks and thickets that shade the banks.

Trouble Brewing

With a noisy splash, **Merten's Water Goannas** drop from overhanging trees into the water at the slightest disturbance. While they are a familiar sight along Top End watercourses, numbers are disturbingly reduced wherever introduced Cane Toads turn up. As the poisonous toads spread along watercourses, predators like the goannas mistake them for tasty frogs. The error is usually fatal for the lizards.

*Female **Gilbert's Dragons** are shades of grey, but males, like this one, have sharp black and white colours.*

Ta-ta Lizard

Gilbert's Dragons ar the most commo and obvious lizard along the edges c Australia's norther rivers. They **bask** on log and branches, and das away on their hind leg when approachec Gilbert's Dragons ofte lash their tails, bob the heads and wave their fror feet. Some people ca them 'Ta-ta Lizard because they look lik they are waving good-bye

Dangerous World

This lone baby **Hawksbill Turtle** is paddling bravely through clear waters. So far it has dodged ghost crabs, gulls and other beach predators to reach the water, but it will be lucky to survive. It must now face the hungry fish. Only one in several thousand turtles will grow up. But like its brothers and sisters, the little turtle is perfectly formed, has keen vision and a good set of flippers to propel it swiftly through the water. This could just be the one!

Tied to Land

Sea turtles remain closely bonded with land. Every two years or more, a female turtle migrates hundreds or thousands of kilometres from her feeding waters to the beach where she hatched. Using her back flippers, she digs a deep burrow with a chamber at the bottom in which to lay her eggs.

*This **Hawksbill Turtle** is laying 100 or more eggs. She will bury them and return to the sea.*

Turtles in Trouble

For millions of years they have roamed the seas. They have dragged their great heavy bodies ashore on remote beaches and provided vital food for people and marine creatures. Now turtles like this **Green Turtle** are in trouble. Chemicals, plastic bags, powerboats, fishing nets and over-catching are reducing their numbers. Scientists are studying sea turtles to find out as much as possible about them.

*The nostrils of this **Elegant Banded Sea Snake** are now closed.*

*This **Spine-bellied Sea Snake** is particularly spiny, which probably means it is a male. Females have smoother skin.*

Closed Nose

Even though **sea snakes** spend their entire lives at sea, they still need to breathe air. In between probing the muddy bottom for eels and searching among the corals for fish, they must swim to the surface for oxygen. To help them live as air-breathing animals in the sea, all sea snakes have nostrils that can open and close. These are located on the tops of their snouts to make breathing easier.

Cast Away

Sea snakes are born in the sea, live and feed in the sea and even give birth to live babies in the water. They never choose to venture onto land, but this one has been caught in currents and dumped on the beach by waves. Its flattened body and paddle-shaped tail are useless for moving when it's out of the water. If it cannot return to the sea it will soon die.

Thin Blue Line

It is a fine line between land and sea, two worlds so utterly different that few animals and plants can survive in both. Most reptiles live only on land, but some live only in the sea. To adapt to **marine** life, limbs have become flippers and tails have turned into paddles. Marine turtles swim in the sea but come back to land to lay their eggs. Keeping warm is a problem, because marine reptiles are largely at the mercy of the ocean temperature. This means that almost all marine reptiles live only in warm **tropical** waters. Shallow seas, clear coral reefs and murky river mouths are home to turtles and sea snakes, and even crocodiles have been found many kilometres from land.

Turtle Alert!

Newly hatched turtles find the sea using the light from the horizon as a guide. If you're camping where turtles are hatching at night, they may be attracted to your lights. So turn off your lights and put out your fires to help the turtles find the sea.

Warm Girls and Cool Boys

The heat inside a turtle nest affects whether the young are males or females. High temperatures produce females and cooler temperatures create males. This means more females will be born on beaches with dark sand that becomes hotter, while more males hatch on beaches with paler, cooler sand.

*It took up to 12 weeks for this young **Loggerhead Turtle** to hatch. It then took a day or so dig its way to the surface.*

All Their Eggs in One Basket

These eggs were laid by several **Garden Skinks.** Up to 50 or more eggs may be stashed under a brick pile or in a compost heap. We do not know why many of the skinks in your yard (plus the ones from next door) sometimes choose to lay their eggs together, but we do know that they may use the same site over a number of years. Maybe they follow each other's scent to a site where egg **incubation** has previously been successful. If you find a **communal** egg clutch in your garden, do not disturb the eggs or they may not hatch.

Each egg is the size of your finger nail.

City Slickers

You don't have to trek through deserts and rainforests to meet reptiles. They live in all of our towns and cities. Two-metre pythons swallow possums and cats in Brisbane and Sydney. Tiger Snakes thrive in some Melbourne suburbs. Brightly coloured burrowing snakes wriggle beneath the sand along Perth's coastal strip. Blue-tongued lizards crack snail shells in backyards across the country. These are the 'city slickers', the lucky reptiles that can adapt to the changes humans have created. For them, compost heaps, piles of old boards and even the gaps above our ceilings and behind cupboards are just as comfortable as many natural **habitats.**

Home Invaders

House geckos have it made! In **tropical** cities like Brisbane, they grow fat on home-delivered food, which arrives each night in the form of insects attracted to lights. There are also plenty of places to hide— behind doors, pictures and wall hangings.

*The **Eastern Blue-tongue** lives in northern and eastern Australia.*

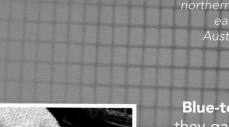

*This **Dtella** is quite happy running up glass. Thousands of tiny bristles on its padded feet create an enormous surface area so that some geckos can walk on surfaces at any angle —even on the ceiling.*

Backyard Glutton

Blue-tongues are well named! When threatened they gape their mouths and display their flat blue tongues to look as fierce as they can. These slow-moving skinks are not fussy eaters. They are happy to dine on fruit, snails, eggs, flowers, dog food or even a litter of baby mice if they can find one. Perhaps that is why they are common in most towns and cities.

Suburban Dragons

Suburban creeks and ponds from Canberra to north Queensland are great places for **Water Dragons**. In some public parks and gardens they can become quite tame. One lizard had to be removed from a hardware store because it developed a habit of lurking under tables and nipping peoples' wriggling toes! By day these large lizards like to laze beside the water, but at night they often lie draped over tree branches.

This tiny young **Water Dragon** *is sleeping on a flyscreen window cover.*

Environmentally Friendly Rat Trap

Silent, secretive 2-metre giants like this **Diamond Python** can live quietly in our roofs without us knowing. That is, until we climb up above the ceiling to discover rafters draped with shed skins. Diamond Pythons are common in many parts of Sydney. They are clean and efficient rat traps that need never be set or emptied.

Heat-seeking Python

Imagine being able to feel the change in temperature when a mouse walks into a room. Pythons can do this! They use special pits along their lips to locate warm-blooded prey by its body heat. The heat travels in a straight line, and casts a shadow inside each pit. This helps tell the pythons where their prey is coming from.

You can easily see the pits along the lower lip of this **Carpet Python.**

pits

Vanishing Resource

Aborigines and Torres Strait Islanders have enjoyed eating turtles, especially **Green Turtles,** for thousands of years. But now, like other sea turtles, they are threatened by pollution, hunting, fishing nets and boat propellers. Because they are a traditional food, **indigenous** people are permitted to hunt turtles. We hope that turtles and their eggs can continue as part of their diets, but the turtles are quickly disappearing. They also migrate, often venturing beyond Australia's shores to countries where they have no protection at all.

Missing!

The only **Retro Sliders** ever photographed are dead specimens preserved in museums. Scientists do not even know if this small burrowing skink still exists. None have been seen since 1960. For years, **herpetologists** have been visiting the only region the lizard has ever been found, the Retro Station and Clermont area in central Queensland, to dig, rake and rummage, without success.

This distinctive lizard, with its forelimbs reduced to dimples and its hind limbs to stumps, may have vanished before we even realised it was in trouble!

Rarest of the Rare

Few other Australian animals are protected by electric fences! **Western Swamp Turtles** live only in tiny winter swamps just north of Perth. In summer, when the swamps dry out, the turtles bury themselves under clay and vegetation but unfortunately foxes dig them up. Turtle numbers dropped to only about 50 during the 1980s. Now, thanks to fox-proof electric fences, fire management and a careful captive breeding and releasing program, there could be as many as 400 Western Swamp Turtles.

Going...Going...

The world is quickly changing. Forests are felled, wetlands filled and woodlands flattened. Expanding cities need more land, fuel and water, and **feral** animals are thriving. Some reptiles actually welcome the new environment—Brown Snakes eat introduced mice in pastures where forests once stood, and some geckos are happier on houses than in trees. But many other **species** are in serious trouble. If we continue

Mountain Retreat

The **Alpine She-oak Skink** is considered to be endangered in Victoria. It lives in the alpine meadows, tussock grasslands and snow gum woodlands of the Australian Alps. These fragile **habitats** are also home to cats, cattle and horses. In recent years bushfires have raged through the region. Because winters are long and cold, plants have a very short growing season. This means damage from trampling and burning takes years to repair, and could be reducing the lizard's habitat.

This lizard is raising its head and flickering its tongue to copy a snake. It might fool some enemies but good naturalists will notice that the tongue is thick and fleshy, not slender and forked.

Brigalow Blues

The **Brigalow Scaly-foot** is a flap-footed lizard that lives in forests of brigalow—a kind of wattle that once extended over broad expanses of inland Queensland and northern New South Wales. Brigalow likes rich, dark soil, most of which has been cleared for farming, leaving pockets that are isolated by open pasture. This means that there are fewer Brigalow Scaly-foots as well as several other reptiles that share its dwindling **habitat**.

No Rocks, No Snakes

It's great to have garden rockeries but be careful where the rocks come from. During winter and spring, **Broad-headed Snakes** live only under flat sandstone slabs resting on bare rock along cliff-tops on Sydney's outskirts. The snakes have all but vanished from accessible areas because the rocks have been taken for suburban gardens. Sadly, snake lovers have also played a part, lifting and breaking rocks while searching for the attractive reptiles. Broad-headed Snakes are now listed as endangered in New South Wales.

Activities

Pick'n'Match

Some are smooth. Others come with prickles. But whether plain, spotted, striped or blotched, most of Australia's reptiles have distinctive colours, patterns and textures that are important clues for identification. And when you zoom in, those watertight skins with their neatly arranged rows of scales make great patterns. Up close they take on a whole new look. Match the patterns with the reptiles by placing the correct number in each box.

The answers are on page 44

1

2

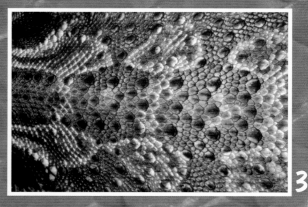

3

Water Dragon (p. 35). Those impressive spines are part of the lizard's crest running along its neck and back.

Golden-tailed Gecko (p. 24). The vivid blaze of colour is a give-away. But why a secretive gecko should have such a bright marking remains a mystery.

Diamond Python (p. 35). The smooth scales overlap neatly and each has its own yellow spot. Compare the picture with the whole snake and see how the 'diamonds' are made of scale clusters with extra pale colour.

4

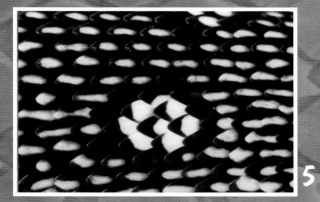

5

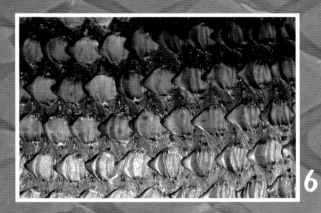

6

7

8

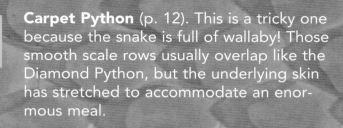

Carpet Python (p. 12). This is a tricky one because the snake is full of wallaby! Those smooth scale rows usually overlap like the Diamond Python, but the underlying skin has stretched to accommodate an enormous meal.

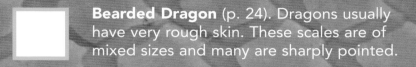

Bearded Dragon (p. 24). Dragons usually have very rough skin. These scales are of mixed sizes and many are sharply pointed.

Thorny Devil (p. 16). A real prickly customer! There are plenty of thorn-like spines while the rich colours—yellow, black, white and brown—are a similar mix to those used in many Aboriginal paintings.

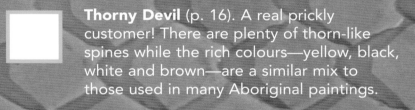

Spotted Tree Monitor (p. 10–11). Notice how the scales are small and arranged side by side, rather than overlapping. Monitors have very tough, flexible skin.

Leaf-tailed Gecko (p. 18). See how the brown, grey and black are mixed and flecked to look like the **lichens** growing on trees and rocks.

39

The Reptile Quiz

How carefully have you been reading? The answers to these questions are all in the book so if you want to find them you'll have to look carefully.

1. Which snakes have legs? (p. 12–13)

2. Why do snakes and goannas have forked tongues? (p. 12–13)

3. What is Australia's largest lizard? (p. 14–15)

4. Which snakes can wriggle their tails to attract food? (p. 16–17)

5. I walk like a clockwork toy and eat up to 5000 ants in a meal. What am I? (p. 16–17)

6. I'm damp, shady and home to lots of reptiles but they are hard to find. What am I? (p. 18–19)

7. Why do rainbow skinks flash their colours and bob their heads? (p. 18–19)

8. There are some places where few if any reptiles can lay eggs, and must give birth to live babies. Where and why? (p. 20–21)

9. Which lizard do scientists believe digs its babies from the nest after the eggs hatch? (p. 24–25)

10. Which snake has enough venom to kill hundreds of thousands of rats in one bite?(p. 26–27)

11. Cook this egg as long as you like, and the white will never go hard. What is it? (p. 36–37)

12. How do baby turtles find the sea? (p. 32–33)

13. What do geckos have under their feet to help them hang on? (p. 34–35)

14. Which reptile do we fear may be extinct? (p. 36–37)

15. Reptiles cannot produce heat so how do they warm their bodies? (p. 8–9)

16. Why does a lizard's broken tail wriggle? (p. 10–11)

17. What lizard adds a squeak to its broken tail? (p. 18–19)

18. Which plant is so attractive to reptiles that some live nowhere else? (p. 44)

Glossary (what words mean)

Ambush Lie in wait to capture prey, rather than actively hunting or chasing.

Aquatic Living in water.

Arid Dry. Most of Australia is desert, receiving less than 250 millimetres of rain per year. Animals and plants must deal with high temperatures and long droughts.

Bask Warm the body by lying in the sun.

Buttresses Support structures growing at the bases of many rainforest trees. Because of these supports, the trees can use shallow roots to gather food near the surface without falling over during storms and heavy rain.

Cloaca The single opening or vent used by reptiles for both reproduction and the passing of wastes.

Communal In one place.

Constriction How some snakes overpower prey by squeezing it in their coils. This is how pythons kill their food.

Cool-temperate Describes the climate in parts of Australia's south-eastern mainland and Tasmania that have cold winters and often snow. Many plants here have small, frost-resistant leaves and reptiles must often hibernate during winter.

Digits Fingers and toes.

Diurnal Active during the day.

Estuary The mouth of a river, where it flows into the sea. Estuaries are often distinct habitats, where fresh water mixes with salt, and mudflats are regularly exposed by the movement of tides.

Eucalypts Gum trees and their relatives. They are common trees throughout Australia, and have been introduced to many other parts of the world.

Feral Introduced. This applies to plants and animals that have been brought to Australia from other countries. When they thrive here at the expense of native species we call them feral.

Festooned Supporting a thick growth. In rainforests, many orchids, ferns, mosses and other plants 'take a short cut' to the sunlight by growing directly on the upper branches of trees, sometimes forming a dense cover that provides a habitat for many animals.

Habitat The type of environment in which an animal lives.

Heaths Shrub communities, usually less than two metres high. Most have hard or prickly leaves. The south-western heaths feature huge numbers of plant species, most of which flower during spring.

Herpetologist	Someone who studies reptiles and amphibians
Incubate	Development of a young animal within the egg. They need warm, constant temperatures, so the mother takes great care to lay eggs in just the right places.
Indigenous	Original inhabitants. This refers to the Aborigines and Torres Strait Islanders who have lived in Australia for tens of thousands of years.
Inhabit	Live in a particular area or habitat.
Larvae	The young of some young insects (such as ants and butterflies) that look quite different from the adults. Grubs and caterpillars are larvae.
Lichens	Flat, plant-like growths on rocks and trees. Some animals that rest on these surfaces have mottled green and brown colours to camouflage them against this background.
Mangroves	Types of trees that grow so close to the coast that the sea covers their roots at high tide and exposes them at low tide. Mangroves are important habitats for some tropical marine reptiles. They help prevent coastal erosion and are breeding sites for many of the fish we eat.
Marine	Of the sea.
Naturalists	People who study nature. If you spend your time watching lizards, snakes, frogs, birds or plants, you are a naturalist.
New Guinea	The whole island to Australia's north, including smaller surrounding islands, the nation of Papua New Guinea and the Indonesian province of West Papua.
Nocturnal	Active during the night.
Non-venomous	Without venom. Many Australian snakes have no venom.
Parched	Dry.
Plague	Rapid population explosion. When conditions are just right, with plenty of grass seeds after good rains, rats and mice sometimes breed in huge numbers.
Population	Members of one kind of animal living in a particular region. It often refers to a group that may be different in some way from others of their own kind, or they may be isolated by water, mountains or desert.
Predator	An animal that captures and eats other animals.
Rodents	Rats and mice. Rodents are not marsupials; they have no pouch and their teeth are designed to gnaw tough foods.
Saplings	Young trees. In rainforests these are often slender and straight as they grow up towards the light.
Scavenge	To forage for scraps and dead things. As well as hunting prey, large goannas often scavenge for animal remains and even discarded food near picnic sites.
Semi-arid	Semi-desert. These areas receive no more than about 500 millimetres of rain per year.

Species

Animals that can breed with each other to produce offspring are called species. Dogs can breed with other dogs but not with cats, and Saltwater Crocodiles can only breed with other Saltwater Crocodiles.

Spinifex

Also called 'Porcupine Grass', these tough grasses with needle-sharp leaves thrive in Australian deserts. The dense clumps offer such ideal shelter that many reptiles live nowhere else.

Tree line

An area on mountains above which no trees can grow. The cold climate on some mountain tops is too severe for trees—only low shrubs and grass tussocks can grow there.

Tropics

Areas north of the Tropic of Capricorn that have mild winters and hot wet summers. The tropics cover all of northern Australia and include the rainforests of the Wet Tropics.

Vegetarian

An animal that only eats plants.

Venom

Liquid injected by some snakes to overpower and sometimes kill their prey. Venom can affect the nerves, blood or tissue. It is injected via the fangs—a pair of hollow or deeply grooved teeth.

Venom glands

Glands on either side of the head that produce and store venom. They are similar to the glands that we have to produce saliva.

Venomous

Having venom. This does not always mean 'dangerous', as many venomous snakes are small, inoffensive and incapable of biting humans. Of course, others are among the most dangerous snakes in the world.

Pick 'n' Match answers

1. Leaf-tailed Gecko
2. Spotted Tree Monitor
3. Bearded Dragon
4. Thorny Devil
5. Diamond Python
6. Carpet Python
7. Golden-tailed Gecko
8. Water Dragon

Want to Know More?

You can find plenty of great information about the animals pictured in this book from other books and websites.

Books

Bennett, R., 1997. *Reptiles and Frogs of the Australian Capital Territory*. National Parks Association of the ACT Inc, Canberra.

Bush, B., Maryan, B., Browne-Cooper, R. and Robinson, D., 1995. *A Guide to the Reptiles and Frogs of the Perth Region*. University of Western Australia Press, Perth.

Cann, J., 1998. *Australian Freshwater Turtles*. Beaumont Publishing, Singapore.

Cogger, H. G., 2000. *Reptiles and Amphibians of Australia*. Reed New Holland, Sydney.

Coventry, A. J. and Robertson, P., 1991. *The Snakes of Victoria. A Guide to their Identification*. Department of Conservation and Environment, East Melbourne.

Ehmann, H., 1992. *Encyclopedia of Australian Animals. Reptiles*. Collins Angus & Robertson, Sydney.

Green, K. & Osborne, W., 1994. *Wildlife of the Australian Snow Country*. Reed Books, Sydney.

Griffiths, K., 1987. *Reptiles of the Sydney Region*. Three Sisters Publications, Winmalee, New South Wales.

Horner, P., 1991. *Skinks of the Northern Territory*. Northern Territory Museum of Arts and Sciences, Darwin.

Holland, S., 2002. *Reptiles DK Eyewonder*. Dorling Kindersley, UK.

Houston, T. & Hutchinson, M., 1998. *Dragon Lizards and Goannas of South Australia*. South Australian Museum.

Jenkins, R. & Bartell, R., 1980. *A Field Guide to Reptiles of the Australian High Country*. Inkata Press, Melbourne.

Pyers, G., 2005. *Life Cycles of Australian Animals. Lace Monitor*. Echidna Books, Melbourne.

Ross, C. A. (Ed), 1989. *Crocodiles and Alligators*. Golden Press, Sydney.

Shine, R., 1991. *Australian Snakes: A Natural History*. Reed New Holland, Sydney.

Slater, P., 1997. *First Field Guide to Australian Frogs and Reptiles*. Steve Parish Publishing.

Slater, P., 1997. *Amazing Facts about Australian Frogs and Reptiles*. Steve Parish Publishing.

Storr, G. M., Smith, L. A. & Johnstone, R. E., 1983. *Lizards of Western Australia II. Dragons and Monitors*. Western Australian Museum, Perth.

Storr, G. M., Smith, L. A. & Johnstone, R. E., 1986. *Snakes of Western Australia*. Western Australian Museum, Perth.

Storr, G. M., Smith, L. A. & Johnstone, R. E., 1990. *Lizards of Western Australia III. Geckos & Pygopods*. Western Australian Museum, Perth.

Storr, G. M., Smith, L. A. & Johnstone, R. E., 1999. *Lizards of Western Australia I. Skinks*. Western Australian Museum, Perth.

Swan, G., 2004. *Green Guide. Snakes and Other Reptiles of Australia*. Reed New Holland, Sydney.

Swan, G., Shea, G. & Sadlier, R., 2004. *A Field Guide to Reptiles of New South Wales*. Reed New Holland, Sydney.

Swan, M. and Watherow, S. 2005. *Snakes, Lizards and Frogs of the Victorian Mallee*. CSIRO Publishing, Melbourne.

Torr, G., 2000. Pythons of Australia. *Australian Natural History Series*. University of New South Wales Press, Sydney.

Weigel, J., 1990. *Australian Reptile Park's Guide to Snakes of South-east Australia*. Australian Reptile Park, Gosford, New South Wales.

Wilson, S., 2003. *Reptiles of the Southern Brigalow Belt*. World Wildlife Fund, Australia.

Wilson, S., 2005. *A Field Guide to Reptiles of Queensland*. Reed New Holland, Sydney.

Wilson, S. and Knowles, D., 1988. *Australia's Reptiles. A Photographic Reference to the Terrestrial Reptiles of Australia*. Collins, Sydney.

Wilson, S. & Swan, G., 2003. *A Complete Guide to Reptiles of Australia*. Reed New Holland, Sydney.

Websites

Australian Museum's Reptile Fact Sheets
http://www.amonline.net.au/factsheets/#reptiles

Herpetological Societies
http://www.jcu.edu.au/school/tbiol/zoology/herp/societies.shtml

Tasmanian Reptiles
http://www.dpiw.tas.gov.au/inter.nsf/WebPages/BHAN-54UVX3?open

Victorian Museum's snake and lizard databases
http://www.museum.vic.gov.au/bioinformatics/lizards/
http://www.museum.vic.gov.au/bioinformatics/snake/

James Cook University's Herpetology Web Site
http://www.jcu.edu.au/school/tbiol/zoology/herp/herp2.shtml

What's in a Name?

Unfortunately the names used for reptiles are not always the same. That's why reptiles and other animals have a special set of names called 'scientific names'. You can use these scientific names to look them up. The reptiles pictured in this book are listed in the Index with both their 'common names' (the names used in this book) and their scientific names.

Index

Acknowledgements

I would like to thank my wife, Marilyn, for sharing the house with snakes and lizards, and for all the times she has been stranded in the hot sun while I indulge my passion. My parents, Ken and Joy, always encouraged me to follow my dreams, which is why Marilyn must sometimes share the house with snakes and lizards and continue to endure the hot sun. Kieran Aland allowed me to photograph some of his specimens. Joy Lawn has given me constructive comments and ideas throughout the writing of this book. During the years I have worked in the Queensland Museum I have met many keen kids with plenty of interesting questions. I thank them all for kindling my own enthusiasm.

Picture credits

New Holland Image Library: Thorny Devil pp. 4, 16; crocodile p. 8, front cover; Green Turtle p. 32, back cover.
Damien Broderick: scientist with turtle p. 9.

First published in Australia in 2006 by Young Reed
an imprint of New Holland Publishers (Australia) Pty Ltd
Sydney • Auckland • London • Cape Town

14 Aquatic Drive Frenchs Forest NSW 2086 Australia
218 Lake Road Northcote Auckland New Zealand
86 Edgware Road London W2 2EA United Kingdom
80 McKenzie Street Cape Town 8001 South Africa

10 9 8 7 6 5 4 3 2

National Library of Australia Cataloguing-in-Publication Data:

Wilson, Stephen K., 1954- .
Snakes, lizards, crocs and turtles of Australia
Bibliography.
Includes index.
For primary school aged children.
ISBN 9781921073014

1. Reptiles - Australia - Juvenile literature. I. Title.

597.90994

Publisher: Martin Ford
Project Editor: Yani Silvana
Educational Consultant: Joy Lawn
Designer: Tania Gomes
Production: Monique Layt
Printer: Tien Wah Press, Malaysia